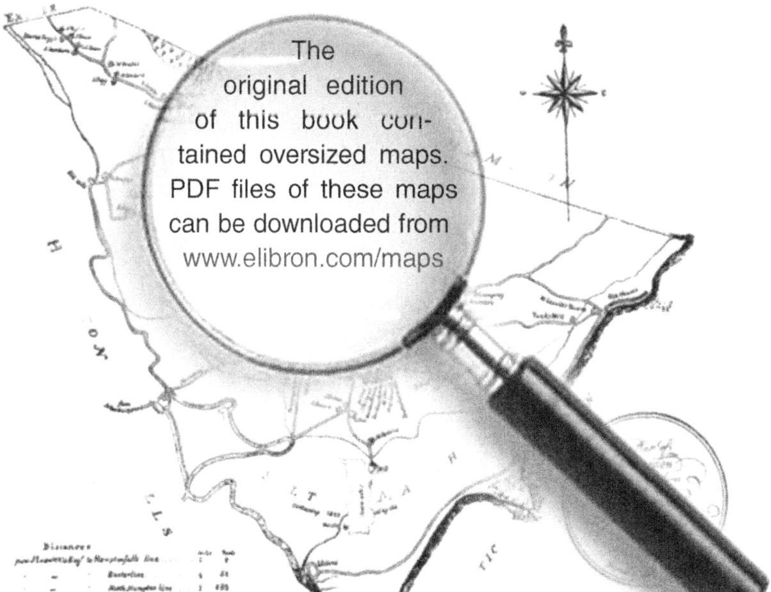
The
original edition
of this book con-
tained oversized maps.
PDF files of these maps
can be downloaded from
www.elibron.com/maps

Joseph de Mendoza y Ríos

On an

Improved Reflecting Circle

Elibron Classics
www.elibron.com

Elibron Classics series.

ISBN 0-543-70442-4 (paperback)
ISBN 0-543-70441-6 (hardcover)

This Elibron Classics Replica Edition is an unabridged facsimile of the edition published in 1801, London.

ON AN

IMPROVED REFLECTING CIRCLE.

BY

JOSEPH DE MENDOZA RIOS, F.R.S.

FROM THE

PHILOSOPHICAL TRANSACTIONS.

LONDON:

PRINTED BY W. BULMER AND CO. CLEVELAND-ROW,
ST. JAMES'S.

1801.

ON AN

IMPROVED REFLECTING CIRCLE.

Read before the ROYAL SOCIETY, *June* 4, 1801.

IN Practical Astronomy large instruments are useful, not only to enable the observer to read the angles to a small fraction of a degree, but likewise to diminish, in the construction, the inaccuracies which proceed both from the errors of the divisions and the eccentricity of the index. Frames of considerable dimensions admit also the application of telescopes with great magnifying powers, which is a circumstance of the utmost importance in celestial observations. As the reflecting instruments employed at sea are supported by the hand, their weight and scale are limited within a narrow compass; and it seemed very difficult to obviate, by any expedient, the inconveniences arising from the smallness of their size, while it was impossible to increase it. The celebrated TOBIAS MAYER contrived, however, a method to determine, at one reading, instead of the simple angle observed, a multiple of the same angle; and, by this means, the instrument became, in practice, capable of any degree of accuracy, as far as regards the above mentioned errors. His invention is essentially different from the mere repetition of the observations; and my object requires that I should explain the principle upon which it is founded.

Mr. MAYER proposed to complete the limb of the Sextant, making a whole Circle, with the horizon glass moveable round the centre, with an additional index, which I shall call *the horizon index*, in order to distinguish it from *the centre index*, to which the centre glass is attached. This instrument is represented in Plate XXIX. Fig. 1; and the manner of using it is as follows. After the index A is set at o, (the beginning of the divisions,) the two glasses are rendered parallel, as is usually practised with HADLEY'S Quadrant, by moving the horizon index B, till the horizon of the sea, (or the sun, or any other object,) or its direct image, and the doubly reflected image of the same, seen through the telescope, coincide. After fixing the horizon index in that position, the centre index A is to be moved, in order to measure the distance of the two objects S and L, (which I shall suppose the sun and moon,) by bringing into contact the doubly reflected image of the sun with the direct image of the moon, seen through the telescope. The centre index will then be at M, and the arch o M might give, as in the Sextant, the angular distance required; but the construction of the Circle renders it easy, in this position, to effect again the parallelism of the glasses, and to make another observation of the contact, in the like manner as from o; which operation will bring the centre index to N. The index will then give o N, or double the distance; and, as it must be divided by 2, in order to have the angle required, the errors of division and eccentricity, which, together, I shall call the *error of the instrument*, will be likewise reduced to one half. It is obvious, that by successive repetitions of the same process, triple, quadruple, &c. the distance may be obtained, and the said error further reduced, in the inverse ratio of the multiplication of the distance, to any degree of approximation required.

The method of rendering the glasses parallel, by means of the horizon of the sea, is not accurate, on account of the indistinctness of the images; and, when the sun is used for that purpose, the observation becomes fatiguing to the eye. The repetition of that operation, by one or the other method, remained therefore a considerable inconvenience attached to Mr. MAYER's Circle. The author himself seems to have been of that opinion, as he proposed to provide the instrument with a diagonal rule, fixed upon one of the indexes, so that the other index should touch it when the glasses were parallel; but an adjustment of this nature must be subject to great errors, and was never adopted in practice. The Chevalier DE BORDA, wishing to remove that imperfection, had the happy idea of rendering the parallelism of the glasses unnecessary, by substituting the observation of the angular distance of the two objects, to that of the coincidence of the images of the same object. This constitutes the second great improvement of the Reflecting Circle, which it is necessary for me to explain, before I proceed to the account of my own investigations.*

In BORDA's Circle, the telescope is fixed at some distance from the centre, and the horizon glass is carried near the border of the instrument, as in Plate XXIX. Fig. 2. By this arrangement, the rays of light can arrive at the centre glass, both from the heavenly bodies situated to the right of the horizon index, as S′, and from those situated to the left, as S. Thus, if the glasses are parallel to one another, when the centre index is at o, it is obvious that there are two ways of making the observations. While the

* It does not belong to my present plan, to explain the effect of BORDA's improvement in correcting the errors which arise from the want of parallelism in the surfaces of the glasses. This will be fully considered in another Paper, where I intend to give an account of several investigations which I have made upon that particular subject.

direct image of the moon L is seen through the telescope, the angular distance to the sun, if at S, may be measured by moving the centre index to *m*, in order to produce the contact; or, if the sun is at S′, the same operation may be performed, by using the contrary motion to *m′*. The first kind of observation, the Chevalier DE BORDA calls *observation to the left;* and the second, *observation to the right.* Suppose now, that (the horizon index being fixed in the same position) the distance from L to S is observed to the left, by bringing into contact the doubly reflected image of S with the image of L, seen without reflection; let us then turn the instrument round, keeping it in the same plane, so as to have the direct image of S through the telescope, and thus make an observation of the same distance to the right; the position of the centre index being in the first observation at *m*, and in the second observation at *m′*, it is clear, that if o is the point where the parallelism of the glasses takes place, o *m* is equal to o *m′*; and, that the arch *m m′*, determined by the two positions, will give double the distance.

It will be more convenient to have the centre index at o, when the first observation is made, in order to take the double distance at one reading, after the second observation. For this purpose, the first part of the process may be inverted, by previously fixing the centre index at the beginning of the divisions, and moving the horizon index H towards o, instead of moving the centre index A to *m*, or towards H.

The last kind of observation, in which the incident ray, which produces the first image upon the centre glass, may be conceived to run double the angular distance, passing in its way over the line of collimation, has been called, by the Chevalier DE BORDA, *the crossed observation.*

The same process may be repeated, by fixing alternately one of the indexes, and moving the other, and continuing successive sets of observations; each set of two crossed observations, one to the right and another to the left. The angle given by the instrument, will be equal to double the angular distance multiplied by the number of sets observed, or, in other terms, to the angular distance multiplied by the number of observations, which are always supposed to be made by pairs; an odd observation being of no value in this manner of using the Circle.

I have expressed myself as if the observations could always be made by looking alternately at each object through the telescope, in order to bring into contact the doubly reflected image of the other object. This is not the case in the observations of the distances from the moon to the sun, or a star; it being then indispensable to compare, by reflection, the brightest of the two heavenly bodies; but there is a very easy method of obviating that inconvenience. After the contact of the images of S and L is observed, with the telescope directed to L, the position of the plane of the instrument may be inverted, turning it round the axis of vision OBL; the incident ray will then answer to the point S′, equally distant from L as S, and the crossed observations will still give SS′, or double the distance.

Whether a Circle is used simply, as Mayer proposed it, or according to Borda's method, its peculiar advantages chiefly depend on the multiplication of the distance required. I have therefore turned my attention to the improvement of this principle; and, with that view, I have contrived the construction which I am going to describe.

In the crossed observations made with Borda's Circle, the indexes move alternately through an arch which, in the divisions, is equal to double the distance: for example, the centre index comes, in the first crossed observation, from m to m'; in the third crossed observation, from m' to m'', &c. and the horizon index, in the second crossed observation, to b'; in the fourth crossed observation, to b'', &c. and, by each of the two indexes may be found the same multiple of the distance required. Let us now place the Nonius in a circle moving round the centre, over, or adjacent to, the usual limb which contains the divisions: it will easily be conceived, that, by attaching that circle, which I shall call the *Flying Nonius*, alternately to each of the indexes, it will serve as Nonius for both; and that, after any number of observations, it will give the compound motion of the two indexes. Thus, after the first observation, the Flying Nonius will, at each crossed observation, advance double the distance over the divisions, while each separate Nonius, fixed on the indexes, requires a set of two observations, to produce the same effect in BORDA's Circle.

Plates XXX. XXXI. and XXXII. exhibit a perspective view, a plan, and a section, of the instrument, which, for the sake of distinction, I shall call my *Improved Reflecting Circle.* The last Plate is particularly intended to shew the compound handle, which I have adapted to the instrument, in order to hold it with convenience and ease in every position.* These three repre-

* The use of MAYER's Circle, or of BORDA's, as constructed till now, with only one handle attached to the centre, is extremely inconvenient in several positions, and particularly when it must be kept inverted downwards during the observation. For this reason, I thought it of importance to contrive such a support as would enable the observer to hold the instrument with the same ease in every direction. This is effected by means of the compound handle, attached to the horizon index, by the brace V and screw X, (Plates XXX. XXXI. and XXXII.) which index turns round the centre with the handle. When Mr. TROUGHTON began to construct this sort of instruments, I

sentations, in which the same parts are marked with the same letters, are sufficient to give an accurate idea of the arrangement of the whole, and make it unnecessary for me to enter now into a minute detail of the mechanism of the apparatus. I therefore shall content myself with adding here only what concerns the general use, and the peculiar properties, of this instrument.

M is the divided limb of the circle, and N the Flying Nonius, (Plates XXX. XXXI. and XXXII.) to each of which the horizon index may be occasionally attached, by means of the clamps D, C; as well as the centre index, by means of the other clamps A, B. The peculiar property of the instrument being that of giving double the distance, I have thought proper to divide the circle into 360 degrees, and not into 720 according to the nature of the Sextant. Thus, after a crossed observation, the reading of the Nonius will, without reduction, exhibit the measure of the simple distance. I have likewise extended the Nonius round the circumference, so that, by the coincidence of two divisions, the number of degrees will appear on the limb, and that of the minutes and seconds on the Flying Nonius. The manner of making the observations with this instrument is as follows.

recommended to him this improvement, which he has adapted to his Reflecting Circle.

I shall observe here, that Mr. TROUGHTON's Circles are not of the kind which I have endeavoured to improve. The scheme of his construction may be said to consist in completing the limb of a Sextant to the whole circumference, and making it capable thereby of performing BORDA's crossed observations, with as many Noniuses as may be attached to the centre glass. But, Mr. TROUGHTON's instrument is deprived of the principle invented by MAYER, for obtaining at one reading a multiple of the distance required; which is the great property of Circles, and, in my opinion, the best means of diminishing their errors.

Adapt the o of the Flying Nonius to 360° of the limb; and then fasten the two clamps A, B, of the centre index* E E, by which the divisions will be kept in the same relative situation. Then, turn round the horizon index F F, and make an observation of the distance to the right. The contact must be adjusted by the screw G, the clamp C being fastened. Leave this clamp fastened, and loosen the clamp A; thus turn the instrument, and make a crossed observation to the left, adjusting by means of the screw H, after having fastened the clamp D. At the end of this observation, the Flying Nonius will give the distance. Fasten now the clamp A, and loosen the clamps B and C, leaving the clamp D fastened; then turn the instrument again, and make a crossed observation to the right. At the end of this observation, the Flying Nonius will give double the distance. By successively inverting the use of the clamps, this alternate process may be continued *ad libitum;* and each crossed observation will increase the reading, by an arch equal to the distance.

Let the number of observations be n, and the angular distance D. The arch given by my improved circle will be $= D(n-1)$. In BORDA's Circle, (reducing the divisions of the Sextant to those of the Theodolite I use,) the arch is $= D \times \frac{1}{2} n$; and, either n must be an even number, or the odd observation must be lost. In MAYER's Circle, the arch is $= \frac{1}{2} D \times \frac{1}{2} n$; and the number n, which comprehends the observations for the parallelism of the glasses and those for the distance, must likewise be even. The

* Properly speaking, neither the centre nor the horizon indexes act, in this instrument, as such, both of them being deprived of the Nonius, which is transferred to the flying circle; but, for the sake of perspicuity, I continue the use of those expressions, in order to distinguish the plates or rules which carry the centre and the horizon glasses.

comparison of these expressions, shews at once the relative advantages of the different instruments.

My construction offers considerable advantages, in every manner of using the Circle. If, instead of the crossed observations, it should be wanted to employ the usual practice of rendering the glasses parallel, a multiple of the distance may still be obtained by my instrument, equal to that of the other method. For this purpose, the parallelism of the glasses may be effected, by means of the images of the sun, or the horizon of the sea, moving the index F, while the o of the Nonius is adapted to 360° of the limb, and the two clamps A, B, are fastened. After this, an observation of the distance to the right may be made, with the clamp A fastened, while the clamp B is loose; the clamp D being also fastened, and the clamp C loose; and, at the end of this observation, the Flying Nonius will give an angle, which will be only the half of the distance in my divisions, but which would be equal to the whole distance, if the divisions were according to the Sextant. After that, and while the clamps B and C are fastened, and the clamps A and D loose, the parallelism of the glasses may be again effected; and the Nonius will advance the same quantity over the limb. The same addition will take place, by inverting the use of the clamps, and making another observation of the distance. The like alternate process may be continued indefinitely; and the result given by the instrument will, with only one observation more, be the same as that of BORDA's method, and double the arch which would be obtained by MAYER's Circle.

Mr. BORDA's Circle is liable to a very great inconvenience in practice. Each index advances successively over the limb; and, in order to facilitate the operation of bringing the images for the contact within the telescope, that author advises to make a preparatory memorandum of the positions which the indexes will nearly occupy, so that they may be set accordingly, previous to each observation. But this method, which is always inconvenient, by night becomes almost impossible. For this reason, I have joined to the horizon index an arch LL, (Plates XXX. and XXXI.) which is divided, both to the right and left, into degrees and minutes of the Sextant, so that, when the glasses are parallel, the centre index coincides with the two first divisions o, o, and occupies the space left blank between them. I have further provided two sliding pieces P, P, which may be adapted to that arch, with a spring sufficient to keep them firm in any situation. Putting each of these pieces upon the arch, so that their ends may coincide with the divisions marking the rough distance to be measured, no more will be required, than to set the centre index alternately against each piece, before the beginning of the successive crossed observation. The clamp may then be fastened, and the remainder of the motion produced by the adjusting screw; as, if necessary, the index will push the sliding piece further, and leave it at the point where the contact was effected.*

The Flying Circle facilitates the use of any number of Noniuses, which may be applied round the whole circumference; but, as the leading principle which I have chiefly had in contemplation, is that of obtaining an accurate result from one reading, I have only used a single Nonius. Two Noniuses,

* The idea of this simple contrivance, was suggested to me by the ingenious Mr. E. TROUGHTON.

opposite one another, might however be advantageous, in order to correct the errors of eccentricity; but, in my opinion, a greater number ought not, in any case, to be used.

Before I conclude this Paper, I shall remark, that my improvement may be partially applied to a Circle, where the telescope and the horizon glass are attached, or fixed, to the main frame of the instrument. The Flying Nonius, acting then with the single centre index, will only give the same result as BORDA's Circle; but this construction seems to me greatly preferable to all the other plans executed till now; the whole apparatus being more solid and simple, and its use not liable to the errors which arise from the motion of the horizon index.

With this construction, we may likewise employ a method of ascertaining the place where the parallelism of the glasses was observed to take place, and of setting the index afterwards in the same situation, as often as is necessary for the repetition of the observations. A piece may be used, so contrived as to be attached to, or detached from, one side of the index, by means of a screw; and provided besides with other screws, to fasten it to any part of the limb. This rectification piece, being previously attached to, and carried with, the index, must be fixed in the situation it occupied when the contact of the images was observed. The index will then be detached from it, in order to observe the distance, and afterwards must be brought back to the same position as before, contiguous to the rectification piece. The like alternate process may be repeated; and the Flying Nonius, going with the index in the motions forwards, and standing still in the motions backwards, will give the multiple of the observed angle, without performing the observation for the parallelism more than once in the beginning. In MAYER's Circle, as well as in BORDA's, there is a great objection to any attempt for that purpose; because, as the horizon glass moves round with the index, its perpendicular position is deranged by the inequalities of the plane of the limb; but, in my construction for multiplying with the horizon glass fixed, that inconvenience is removed; and the method of ascertaining the identical position of the glasses may be employed in practice, with advantage, it being done when the index is at the same point of the frame. By suggesting this idea, I do not, however, mean to represent it as preferable to the repetition of the observations; which process must, for many reasons, have the advantage over any mechanical contrivance whatsoever.

I have procured Reflecting Circles to be constructed, upon the principles here described, both with the telescope and the horizon glass upon a moveable index, and with the same pieces attached to the main frame of the instrument. The two methods have respectively answered my expectation; and I purpose, at a future opportunity, to publish a description of the means which I wish to recommend for the mechanical improvement of the different parts, together with an account of some other essays which I have made relative to the same subject.

Printed by W. Bulmer and Co.
Cleveland Row, St. James's.

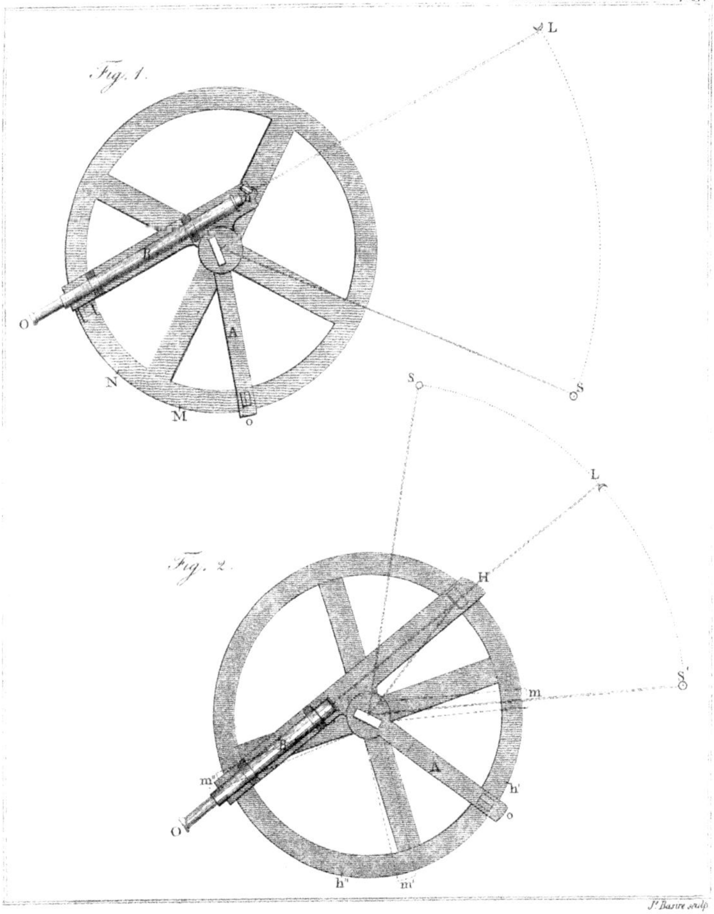

J. Basire sculp.

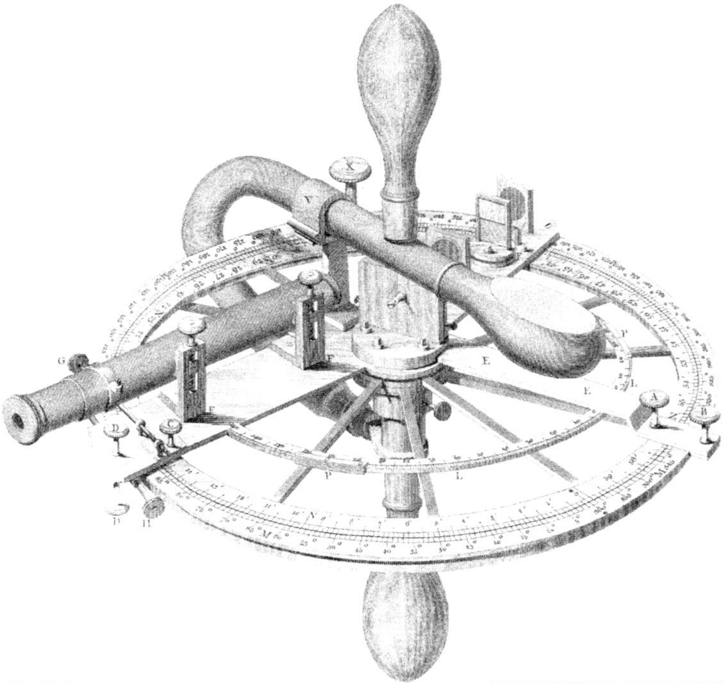

J. Basire sculp.

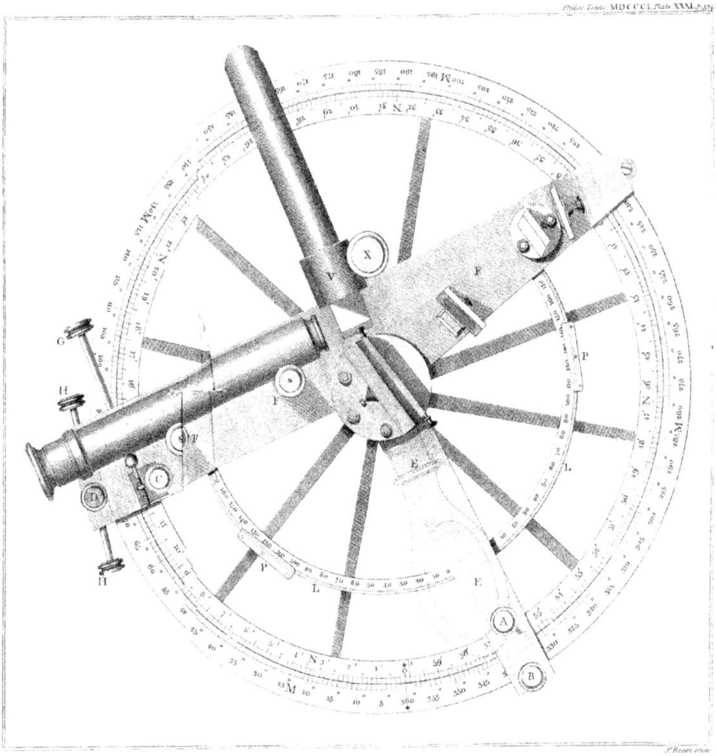
X
V
F
P
G
H
H
D
C
E
E
L
L
P
A
B
M
N
M
N

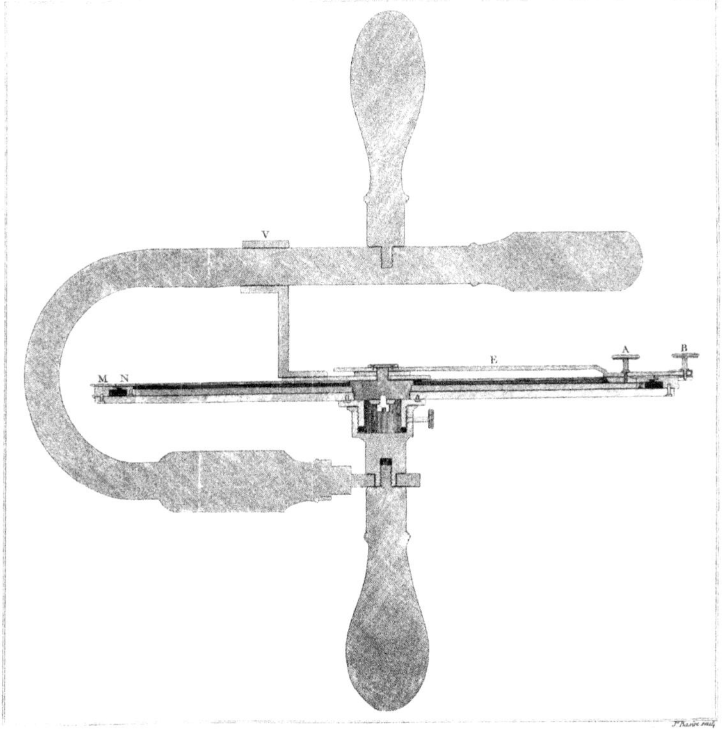

J. Basire sculp.

Printed in Dunstable, United Kingdom

91220873R00020